THE SOURCE OF EVERYTHING

The Hidden Truth

———❊❊❊———

DAVID THOMAS

WestBow Press books may be ordered through booksellers or by contacting:

WestBow Press
A Division of Thomas Nelson & Zondervan
1663 Liberty Drive
Bloomington, IN 47403
www.westbowpress.com
1 (866) 928-1240

ISBN: 978-1-4908-6634-5 (sc)
ISBN: 978-1-4908-6633-8 (e)

Library of Congress Control Number: 2015901679

Printed in the United States of America.

WestBow Press rev. date: 02/06/2015

WESTBOW
PRESS
A DIVISION OF THOMAS NELSON
& ZONDERVAN

Table of Contents

Fact and image credits are presented at the end of the novel.

Prologue

Often, you find yourself asking the same questions - Should I wear this shirt today or that one? Should I really be eating this calorie-packed donut? What am I doing with my life? In fact, just what is life? Oh, and where did it come from? Wait – that last one... Where did life come from? Seriously, where did everyone, and everything you can think of, originate? Well, there are many ideas on how life (and the universe itself) came into being, but two of the most prevalent theories of them all are the Creation ideas of the Bible, and the more secular theories such as the Big Bang. Many will say that the the Bible is true. On the other hand, many others will say that "science" is true (referring to the Theory of Evolution, or the Big Bang). Many people support both sides, but which one is really true? A 3,000 year old book or new equipment which allows us to understand the world around us? That is quite the question, so grab a seat, and let's think about it.

Before I tell you which one is correct and why, we ought to discuss what these two theories are. The first one (the Bible story) describes a Creator forming planet Earth, all things which inhabit it, and the rest of the universe. This happened in six days, approximately 6,000 years ago. The latter theory (the Big Bang) states that in fractions of a millisecond, parts of the (what would be) universe suddenly "exploded" and then started to expand outwards, creating what we know as the universe today. Afterwards, countless clouds of gas, dust, and other sorts of debris swirled around

This image displays how planetesimals would swirl around the growing star. The clumps of gas and dust would then be turned into planets and moons.

itself. The gas and dust collapsed upon itself in the center of a cloud, which would form stars, and the rest of the debris that was circling around in that cloud would then start to clump together, forming the planets, their moons, and the asteroids within that particular solar system (this debris is known as planetesimals.) This is how our solar system came to be, too, about 4.6 billion years ago. Then after a period of time, life came to our planet, and here we are today.

Our Solar System

Our solar system is confirmed to consist of one star and eight planets. Viewing the image from left to right, you can see the Sun, Mercury, Venus, Earth, Mars, Jupiter, Saturn, Uranus and Neptune. Our solar system also contains comets and dwarf planets. The distances and sizes of the objects shown in this image are not to scale.

There is more to the processes of both theories than that, but we will focus on the astronomical facts behind them. Of course, the Bible makes it clear that a Creator formed the Earth, and then the job was done. It is difficult to find evidence for such an event, so we will see whether or not cosmology is correct (cosmology referring to the secular Big Bang and nebula theories). We will first discuss our solar system and the objects within it, and then move outwards and discuss stars.

However, before we even discuss those objects, there is already something to know about this cosmological model! Remember how the planetesimals would have stuck together to form the planets? Well, once dust clumps get to a certain size, instead of sticking together with other clumps, they impact each other too rapidly, and destroy each other (gravity would not be able to overcome this until after they have formed into larger bodies.) Astronomers have even witnessed such an event occur! One astronomy textbook says-

"Once these planetesimals have been formed, further growth of planets may occur through their gravitational accretion into large bodies. Just how that takes place is not understood." - Martin Harwit, *Astrophysical Concepts*, 2nd ED., p.553

The Sun

One flaw is not necessarily enough to debunk an entire theory. Anyways, what does the majestic Sun have to say about such a theory? Wait - what is the Sun? The Sun is a star, but it is more than that. It is a somewhat unique star, and its unique features are required for life on our home planet, Earth. For example, the Sun is a type G star. Unlike most stars which range from type O to M, our type G star does not produce harmful radiation. Plus, it has the perfect temperature for us, and it has a relatively "quiet" surface (which means that harmful solar flares do not occur often). If it did not, it would be impossible for life to exist.

"Sun-like stars normally produce a bright superflare about once a century....Why a superflare has not occurred on the Sun in recorded history is unclear. 'I think a consensus is emerging that our Sun is extraordinarily stable', suggests Galen Gisler, an astronomer at the Los Alamos National Laboratory in New Mexico." - "Thank our Lucky Star", *New Scientist*, 161(2168) 15, 1999

"Our Sun, it seems, is favored with anomalous stability, but no one knows why. We are simply lucky!" - http://www.science-frontiers.com/ sf122/sf122p03.htm

This flare is one that is found from our Sun. Ones from other stars can be over one hundred million times more energetic, which would wipe out all life on Earth.

Or maybe it was not just by luck.

Also, our Sun is just the correct size for life on Earth, and is the perfect distance away. Not to mention, the majority of stars are binary, meaning two or more stars orbit around each other. If our Sun was a binary star, we would basically have two suns, which would undoubtedly make our planet uninhabitable.

However, there is more to the Sun than just things that could have potentially happened by chance. One problem is that as stars last longer, they grow larger. If our solar system is 4.6 billion years old, as the secular theories state, scientists have calculated that our Sun would have been about 25% smaller in the past. That may not sound too important, but that would have been enough to prevent life from evolving on Earth! Our oceans would not have been heated enough, and heat in the oceans is required for most evolutionary theories on Earth to prove valid.

Another serious problem for cosmology is that the 8 planets in our solar system (Mercury, Venus, Earth, Mars, Jupiter, Saturn, Uranus, and Neptune) do not orbit around the Sun's equator. This is a huge problem because the gas cloud that would have formed our solar system would have needed to condense at the Sun's equator, and then the planets would orbit around the newly formed star in that plane. However, the 7° tilt that we see today shows that the cloud could not have condensed where it was suspected to, and if the cloud did not condense at the equator, then there is no area which it could have! Therefore, one can conclude that a cloud of gas and dust could not have formed our solar system.

The Sun also shows no evidence for 4.6 billion years of age, either, as the secular model implies. One secular astronomer said-

"I suspect… that the Sun is 4.5 billion years old. However, given some new and unexpected results to the contrary, and some time for frantic recalculations and theoretical readjustment, I suspect that we could live with Bishop Ussher's value for the age of the earth and sun [about 6,000 years]. I don't think there is much in the way of observational evidence to conflict with that." - John Eddy, quoted by Kazminn, R.G. "It's About Time: 4.5 billion years" *Geotimes* 23, 18-20 1978

The Sun as seen from the ISS. It truly is remarkable how such a star could have all the perfect conditions to sustain life on our planet.

There is obviously much about the Sun that is truly remarkable and cannot be explained. The next time you hear about the Sun proving any theory, just remember these basic facts that were discovered many years ago. The Bible may be vague by only saying that the Sun was created, but that is what the intriguing thing is. If there was no mystery, there would be no point in science. So instead of looking at the Sun and finding it as an object of controversy, remember that it seems to have been created just for us, rather than have formed all by itself.

Mercury

This is an image of Mercury, the closest planet to the Sun.

Mercury is the closest planet to the Sun. It is also the smallest planet. In fact, it is smaller than several moons in our solar system, and just a tad larger than ours. Oddly enough, we humans do not know very much about this planet. Sending one spacecraft there takes as many resources as sending a spacecraft to the outer reaches of the solar system. However, a few spacecrafts have been sent to this small planet of extremes. The first one that was ever sent had something unusual happen to it – Mercury started gravitationally "tugging" on it. This means that Mercury must be very dense inside. That is why most scientists believe that around 85% of Mercury's interior is composed of iron, which means that there is a lot of material packed into a small amount of space! This is very difficult to account for in secular models, because all of the terrestrial planets (the 4 rocky planets near the Sun) should have formed with lighter material on its exterior. Because of this, cosmologists resort to the "large-impact hypothesis". The large-impact hypothesis basically claims that an asteroid or planetesimal slammed into certain planets. This would mean that the light material, which is what would have been required for Mercury to have formed according to the secular model, would have been blown away from the planet. Unfortunately for the secular model, there is no evidence for such a collision like this.

Another problem with Mercury is that it has a magnetic field. A magnetic field (an area in space surrounding an electric force within a planet) does

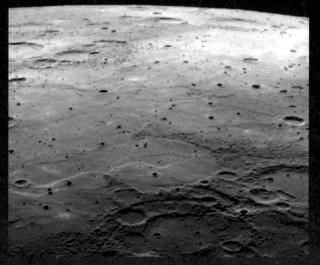

At the very center of the sphere of a planet, there is a core. In order for cosmologists to account for Mercury's magnetic field, they must assume that the core has particular elements already built within it.

several things, including deflecting certain particles away from the planet. "So why does that matter?" you may ask. Well, in order for a planet to have a magnetic field, it must be fairly young. This is because magnetic fields decay, and rather quickly at that. The Earth's magnetic field has actually decreased by about 50% ever since we could measure it. Because cosmology requires planets to be very old, this proves to be bad news for secular theorists. Though, there is a way for a magnetic field to last for a long period of time. It is allowed by something called a dynamo. Simply put, a dynamo is liquid metal flowing within each planet. This produces an electrical current, which then proceeds to produce a magnetic field. However, one astronomy textbook points out-

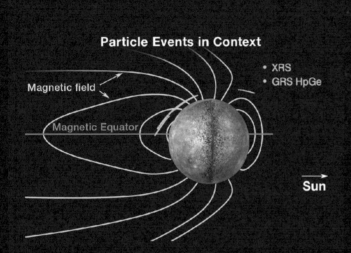

Mercury's magnetic field, as seen above, should not be there if Mercury is billions of years old, and just like Mercury's dense interior, it requires much speculation to accommodate the cosmological model.

"Mercury is so small that the general opinion is that the planet should have frozen solid eons ago [ceasing any electrical current within]." - S.R. Taylor, *Destiny or Chance: Our solar system and its place in the cosmos* p. 163

There is one famous belief that there is sulfur mixed within the iron in the core of Mercury, which would prevent Mercury from freezing solid, and allow it to keep its magnetic field. However, volatile elements such as sulfur could not have condensed so close to the Sun, or else it would have melted away. Despite this, some scientists suppose that maybe the planetesimals that formed Mercury (which is impossible, if you recall) had sulfur already within them. However, there is no evidence for that, and may very well be impossible.

As if all of that were not enough for cosmological theories to handle, recently there have been hollows discovered on the surface of Mercury that seem to be relatively young. This is because these hollows have not gained small craters, as would seem required if they were very old. As space debris strikes an object, it creates large craters, which then causes excess materials to spray up, land, and form many small craters when it comes back down. It is impossible for Mercury to still be "sheltering" these hollows, as well, as all activity relating to the formation of Mercury's surface should not have occurred for all 4.6 billions of years that Mercury is believed to have existed.

"The presence of actively-decaying volatile deposits means the craters cannot be millions of years old, because such geological activity would have ceased eons ago, hence the perplexity of secular planetologists...." - http://creationrevolution.com/mercury-marks-youth/

As we conclude our look at Mercury, there is one word to explain the unexplainable about this mysterious world: speculation. There is no evidence for a collision, and all odds point to Mercury not being able to have a magnetic field, either. Not to mention, the hollows on Mercury cannot be accounted for. Maybe, just maybe, Mercury is not just another space rock that is there just by chance. It could very well be a "messenger" of God, as its name (somewhat) implies in Greek mythology. Though we know comparatively little about this planet, it still makes secular theories appear impractical.

Venus

Venus is often mistakenly considered a bright star. Indeed, this bright world shines radiantly in the evening sky. Some people consider it the "sister planet" of our home Earth, because both

planets are about the same size, and made up of the same materials. In fact, this has led some scientists to believe that Venus and Earth formed at the same time, and at the same place. However, some discoveries have shown that these 2 planets are fairly different. Venus has an atmosphere consisting of CO_2, while Earth's contains nitrogen with some oxygen. Not only that, but Venus has about 30 atmospheres worth of pressure bearing down on its surface. It also has no moon. The Earth and Venus apparently are not that similar after all.

This image of Venus is an ultraviolet image of Venus' clouds.

However, there is something that makes Venus very different compared to the other planets in the solar system. Venus' rotation is retrograde! That means that instead of the Sun rising from the east and setting in the west (like on Earth,) on Venus the Sun rises from the west and sets in the east (basically, Venus rotates in the direction that the other planets do not.) If the secular theory is true, then all the planets should be rotating in the same direction! If you spin a steering wheel (and assuming that the steering wheel will continue spinning forever), it will not randomly stop and spin the other way! One astronomy website states-

"Of all the planets in the solar system, Venus has a unique rotation. Seen from above, all of the planets rotate in a counter-clockwise direction. And this is what you would expect if all the planets formed from the same planetary nebula [the cosmological theory] billions of years ago." - http://www.universetoday.com/14299/retrograde-rotation-of-venus/

So how do secular astronomers explain this? Apparently, there are two theories. One states that the thick atmosphere of Venus slowed down its rotation, but many agree that that one seems too speculative. The other states that an asteroid crashed into it, and the impact was so powerful that it caused the planet to change its direction. However, even if an asteroid could pierce through Venus' thick atmosphere (and still maintain its size and composition to do such damage), there is (just like Mercury) no evidence for such a collision. In fact, Venus' surface is one of the smoothest in the solar system. If there really was such a powerful impact, surely it would have left behind some sort of evidence!

You may be wondering, then, how cosmologists fill this hole in their story. Well, the popular theory is that, about 500 million years ago, the many volcanoes on Venus suddenly erupted, and the magma from these volcanoes sunk into the craters of Venus, efficiently covering up the markings the asteroid left behind. If this sounds like nonsense to you, that is because, well... it is! What would have caused all of these volcanoes to spontaneously erupt like this, and why 500 million years ago? The only answer cosmologists can give to such questions is, "Well... If this did not happen, then our theories would be wrong!"

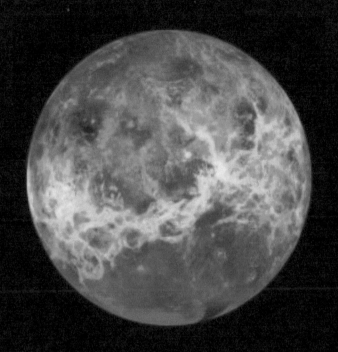

The surface of Venus is very smooth (as you can see). Therefore, it is obvious that no catastrophic collision occurred. The only explanation for this anomaly has no evidence, and it seems unlikely that such an event could occur.

Venus is a perfect example of some of the absurd ideas that cosmologists will believe, just so their theories appear correct. To be fair, maybe these events really did happen, but there is no evidence. Does it not seem odd to you how such wild theories are accepted as fact in the world of science? Science is defined as knowledge gained by **observation** and **experiment**. Nowhere is it said to be opinion! Yet, this is what is being accepted as scientific fact today.

"For those who exalt themselves will be humbled, and those who humble themselves will be exalted." - Matthew 23:12 NLT

Earth

We really do not appreciate how beautiful our home is. Earth truly is humbling. There are massive, sand-filled deserts, snow-blanketed tundras shrouded in auroras, life-teeming green forests, breath-taking coral reefs and so much more that it would take too long to write them all here. Not only is Earth beautiful, but it also has a rich history. Nations have risen in all corners of the world, and many empires have fallen. Countless human lives have breathed oxygen into their lungs, and countless human lives have passed away from this world. Each of these humans had their own feelings, emotions, goals, and desires. What some of these humans have done has affected us today, whether these things were good or bad. But that is enough on that matter.

This beautiful image of our planet makes us realize how little all of us truly are, and how we all share this planet together.

So, how long has our home been orbiting around the Sun? A secular scientist will tell you about 4.6 billion years, just like every other planet in our solar system, but is that really the case? Well, first of all, our magnetic field has been decaying, just like Mercury's. As stated in the Mercury section, it has decayed by about 50% ever since we have been able to measure it, only several hundred years ago. Even if there was a polarity flip in the past, Earth would still only have its magnetic field for about 10,000-20,000 years at most. It is also important to consider carbon dating. Carbon dating is used to determine how old an object is. Carbon dating is interesting because some fossils have been proven to be only thousands of years old, as the Bible suggests. Also, when Biblical archaeologists used carbon dating on Biblical objects to see how old those objects really were (and when the dates were correspondent to dates in the Bible), it had seemed as if carbon dating proved Biblical events. Often, atheists mock at these claims,

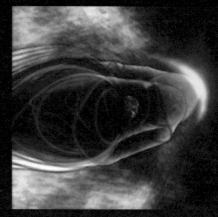

This is an artist's conception of Earth's magnetic field deflecting particles. Not only does it shield us from harmful rays from the Sun, but it also is a perfect example of how our planet could not have been around for so long.

saying that some other elements may have been mixed during the dating, thus making the results inaccurate. Maybe so, but for the amount of objects found, it seems unlikely that all of them were flawed. It is true that some fossils appear to be fairly old, too, but just like some say happened to Bible-endorsing artifacts/fossils, other elements may have been mixed up within those objects, making their records inaccurate.

Another problem for cosmology is that Earth has an abundance of water. Almost 70% of the Earth's surface is water, and the amount of water grants the Earth nicknames such as "The Blue Marble". The reason water is such an issue is because volatile gases such as water cannot have condensed so close to the Sun! So, how do secular theorists solve this? Well, an old theory suggested that back when the Earth was still developing, many comets slammed into Earth. Since comets are considered "dirty snowballs" (and because this "snow" would have probably contained water) the water of the comets would have spewed onto the Earth once they hit the surface of our planet, leaving us with 70% of our planet covered in oceans. Unfortunately for secular scientists, this idea does not work very well anymore. Rovers have been sent to collect materials from comets, and comets apparently contain a large amount of deuterium. Since deuterium is very rare on Earth, our water surely could not have come from comets. The only other explanation for water suggests that asteroids from way out in our solar system (containing water) slammed into Earth. Though this is technically possible, there are many problems with this theory. First of all, an innumerable amount of "water-asteroids" would have probably been required to account for all of the water on Earth. Another problem is, well... how the water would have gotten into such rocks (if the water was not protected deep within the rock, it would have been fried away due to the Sun's distance from the Earth.) The asteroid would also need to have enough amino acids to make it through the Earth's atmosphere, at least to a certain point where the water could be trapped within Earth's atmosphere. The worst part is that, once again, there is no evidence for such an event. It is also important to note that much (if not all) of the Earth's surface would have needed to be covered in water by about a billion years or so after the Earth had formed, or else it would violate theories that involve evolution within the waters.

This image taken from the ISS shows us just how much water covers Earth's surface. And its existence on our planet cannot be accounted for!

"There's one thing on which most geochemists and astronomers agree: The celestial pantry is now empty of a key ingredient in the recipe for Earth." - *Science News*, March 23, 2002

Honestly, though, what are the odds of life for us? Our solar system is placed perfectly between 2 arms of the Milky Way galaxy (areas with much radioactive activity, which would obviously not be good for life on our planet). Also, Earth orbits with a 23.5° tilt. This allows Earth to have seasons. Earth also spins (or rotates) at the perfect speed. If it rotated too slowly, the climate would experience drastic changes. However, if it rotated too quickly, we would suffer from devastating winds. In addition, our atmosphere is composed of the perfect materials for life, and the gases within it that we cannot use are conveniently broken down for us by other natural processes or creatures (for instance, nitrogen is broken down via nitrogen fixation for creatures to biosynthesize, which is crucial to life.) What are the odds for such a balance?

The odds for life on our planet are so slim, we should not even exist!

The human body (and the bodies of all creatures, for that matter) is truly fascinating. Billions of cells are within each body, each very complex, with each working like a factory. The brain's ability to store information is marvelous, and DNA (deoxyribonucleic acid) within us acts as a "blueprint for life." Not to mention, the miracle of childbirth is truly remarkable.

"For the Lord is God, and He created the heavens and earth and put everything in place. He made the world to be lived in, not to be a place of empty chaos. 'I am the Lord,' He says, 'and there is no other.'" - Isaiah 45:18 NLT

The Moon

Also, Planet Earth has one natural satellite. This satellite has intrigued many, and has been an object of worship for many ancient peoples. Yes, that object is the moon. The moon is often treated like the Earth. It is gorgeous, yet we often overlook it. However, despite our ignorance, it is very important to us. It creates a "tug-of-war" with the Earth (due to both bodies gravitationally pulling on each other) which allows the ocean to have tides. These tides create circulation within our oceans, which helps keep our waters clean, allowing marine life to exist. The moon's pull also helps keep Earth somewhat stable as it travels through space, and helps the Earth maintain a stable climate with seasons. And even though the moon reflects about 7-12% of the sunlight it receives back to Earth, it still is our main source of light during the night (when it is during its night phases, of course). According to the Bible, the moon also was made for marking signs, too. Lunar eclipses (or red/blood moons) often symbolize a message for God's people, and solar eclipses (when the moon aligns with the

We have all seen the moon many times. Not only is it beautiful, but it is essential to life as we know it!

daytime Sun and blocks out its light) symbolize a message for the entire world. In speaking of which, our moon is the only moon in the solar system that can cause an actual eclipse!

Lunar Eclipse

Other moons in the solar system are either too small to fully block out the sun, or are too large (because of their parent-planet being too far away) and completely conceal the Sun. Only our moon has the exact size and is the exact distance away to allow such eclipses!

Solar Eclipse

Though, a cosmologist will tell you that all of this happened by chance. A cosmologist will tell you that the moon got here all by itself, too. A cosmologist will tell you these things, yet in reality they do not know how it actually got here. Right now, there are two theories for its existence. One states that at one stitch in time, the moon flew too close to the Earth, and got captured in its orbit. This is impossible (unless there were gravitational interactions with other objects) because the moon would have accelerated out of Earth's orbit. The other (more popular) theory suggests that a Mars-sized object slammed into the Earth. The debris caused by this would condense together and form the moon, and the rest would come back down to the Earth. However, you would need an object of just the right size, striking the Earth at the perfect angle, for this to even work!

"The collision has to be implausibly gentle. You practically need someone to hold a Mars-sized object just above Earth and drop it, to avoid messing up Earth's orbit." - Peter Noerdlinger, Ph.D., quoted in *New Scientist*, 23 January 2007, page 16

Also, after analyzing lunar rocks and sending a rocket into the moon, scientists recently confirmed that water is within the moon! Obviously, the water would have been vaporized during such a catastrophic event.

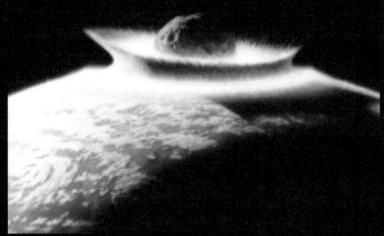

"It's hard to imagine a scenario in which a giant impact melts completely, the moon, and at the same time allows it to hold on to its water…that's a really, really difficult knot to untie." - Erik

This is what a collision that would have produced the moon would look like. Though, there are many problems towards whether or not this was even possible.

Hauri, a geochemist at the Carnegie Institution's Department of Terrestrial Magnetism in Washington, D.C., as quoted in http://www.npr.org/templates/story/story.php?storyId=92383117

The moon is undoubtedly one of the most amazing things for us. Not only is it necessary for life on our planet, and not only is it used for signs, but it is used to show that the Enemy of the Creator (also known as Satan, the Prince of Lies) is truly bankrupt. Planet Earth, and its beautiful satellite, are gifts from the Lord.

Mars

The red planet is the most mentioned planet of them all in social media. For hundreds of years, people have speculated that there may be life on Mars. Indeed, even though we have sent probes to Mars since the 1970's, many people (including astronomers) believe that Mars has life on it, or at least that it may have housed life in the past. The reason many cosmologists want Mars to have life on it is because it would help explain how life arose on Earth (interesting how they tell you that they know how life came to be on Earth, yet in reality they are still trying to find an actual source for life.) They say that there is (or used to be) life on Mars because there seems to be evidence for water on the planet, and water is required for life as we know it.

This is a close up image at Mars red-hued surface.

For instance, there are some canyons that seem as if there was water in them at one point, and there are craters with sediment layers within them. There seems to also be water within Mars' crust, and even water vapor within the atmosphere. Also, by its north and south poles, there are traces of water (though it also contains dry ice and Co2).

So far, this is sounding great for a secular scientist trying to explain life without a Creator. Though, right now water in its liquid form is impossible on Mars, thanks to the thin atmosphere the planet has. That is why secular scientists theorize that at one point, Mars did have a thick atmosphere that could contain water (like Earth's), but then a huge asteroid slammed into the planet, stripping the atmosphere away.

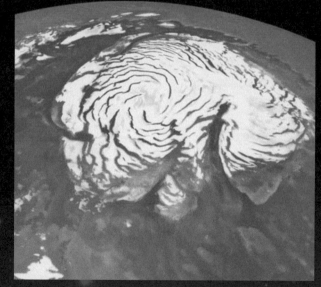

This is an ice cap at Mars' north pole, which could contain water.

However, even if Mars had… Wait. Have you noticed a pattern? In each of the 4 planets we have discussed, secular astronomers have used asteroid collisions to fill in the gaps of their Creation stories… asteroid collisions that we have no evidence for. Anyways, even if Mars had liquid water on its surface, it does not matter in the end. Yes, water is needed for life to exist, but so are other things, too.

Valleys like this one do seem to have once been formed by water (though this particular image is a computer simulation.)

"More than 30 years of experimentation on the origin of life in the fields of chemical and molecular evolution have led to a better perception of the immensity of the problem of the origin of life on Earth rather than to its solution. At present, all discussions of principal theories and experiments in the field either end in stalemate, or a confession of ignorance." - Professor Dr. Klaus Dose, "The Origin of Life: More Questions than Answers," Interdisciplinary Science Reviews, vol. 13, no.4, pp. 348-356

"It is extremely improbable that proteins and nucleic acids, both of which are structurally complex, arose spontaneously in the same place at the same time. Yet it also seems impossible to have one without the other. And so, at first glance, one might have to conclude that life could never, in fact, have originated by chemical means." - Leslie E. Orgel, "The Origin of Life on the Earth," *Scientific American*, vol .21, October 1994, pp. 77-83

It is also important to note that not all terrain on Mars could have come just from water. The volcano eruptions and global sandstorms have had significant impacts, as well.

In addition to the idea of a speculative asteroid impact, many people call even the slightest distinctions on Mars "signs of life." For example, a hi-tech rover took a picture of a rock shaped like a jelly donut. For several days, social media stated that the rock was proof for life on Mars. Some even stated it was life (even though we have discussed how liquid water is physically impossible today.) Soon enough, however, it was discovered that the infamous jelly donut was really just an overturned rock that the rover flipped. Obviously, it made social media, and even some scientists, look really poor. In fact, this particular rock has twice the amount of manganese than any other rock that has been examined, which upsets cosmological theories on how Mars' geological activity is supposed to be. Anyways, another example was when an unanticipated beam of light appeared. Again, some acted as if we ignorant humans know everything, until it was clarified that the beam was just a cosmic light.

As you can see, not everything is just fact. Just because people want there to be life on Mars, does not mean there is. It is important to separate bias from science, both from the Biblical and evolutionary standpoints. Though, if life never existed on Mars, cosmology has to look for another source of life, in which it has yet to find. In fact, the more astronomers set their sights out in the universe and discover other celestial bodies, it confirms the fact that the chances for life are so slim, and that our home is truly unique.

"For there are many who rebel against right teaching; they engage in useless talk and deceive people." - Titus 1:10a NLT

This is a before and after image of the "jelly donut's" location.

The red planet is almost like a red herring, because instead of a "once life-infested planet" as the media says, it really is a planet of speculation. It is a shame that even the planets are not held sacred anymore, and are just used for popularity points.

Jupiter

Just beyond the 4 terrestrial planets, we come to our first gas giant. Jupiter is considered the "King of the Planets" because it has a larger mass than all the other planets within our solar system combined! And just like all of the other planets, Jupiter poses some very interesting problems for cosmologcal models.

This giant planet is very beautiful. It has bands of clouds wrapping around its entire surface, and has dynamic storms, such as the famous great red spot.

Jupiter's "physical features" are very strange. Of course, Jupiter is a gas giant, thus having no actual surface until you go way down into the atmosphere (if you consider the core a surface). However, Jupiter's atmosphere has a very odd chemical composition. This is important to note because the secular theories suggest what chemical properties each planet should have, but Jupiter crushes those predictions! Jupiter is abundant in nitrogen, xenon, and other elements it should not have according to cosmology. Plus, in order for the secular model to work, the core of Jupiter must be a certain size, about 3 times the mass of Earth. When a space probe was dropped into Jupiter's atmosphere, the core was not even discovered! Yes, the probe did only breach through about .2% of the atmosphere, but it still leaves many questions.

"While various spacecraft photographed and collected data about Jupiter since the 1970's, our knowledge of what lies under the atmosphere is very indirect." - http://www.onr.navy.mil/focus/spacesciences/solarsystem/jupiter1.htm

Also, to the surprise of many, Jupiter cannot possibly exist if the secular model was true. This is because the gas cloud that would have formed the new Sun and the planets should have dispersed in around 5 million years. However, scientists calculate that a planet as large as Jupiter would need anywhere from 10-100 million years to form! As I am sure you can see for yourself, Jupiter should not exist at all!

The Galilean Moons of Jupiter

Much like Earth's moon, Jupiter has a fascinating satellite, as well. Actually, Jupiter has many natural satellites, but the majority of them are small asteroids. However, 4 of them are much larger, and have several very unique features. They are called the "Galilean moons of Jupiter," named after Galileo because he discovered them in 1610. These 4 moons are named Ganymede, Europa, Callisto, and Io.

Ganymede

Ganymede is the largest moon in the solar system. It has very dramatic terrain changes all throughout it, and it calls into question some ideas of how such terrain can form. A celestial body should have relatively similar terrain composition throughout.

It is crazy to think that, according to glorified cosmological theories, the largest planet in our solar system should not even exist! What's more, its Galilean moons do not seem to help their ideas at all.

The larger problem about Ganymede, though, is that this moon has a magnetic field! We have discussed how planets can produce their own magnetic field, but the chances for a moon to have one are much slimmer. It also seems unlikely that just the core of a moon alone can keep a dynamo for billions of years.

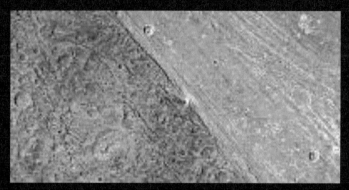
This image of Ganymede's surface shows seemingly fresh and young terrain.

"The heat flux out of Ganymede's core should not be enough to keep convection operating in the core for over four billion years" - as quoted in Ganymede magnetic moon – creation. com http://creation.com/ganymede-magnetic-moon

Who knew that a moon could have a magnetic field!

Europa

Europa is a moon that is given the same treatment as Mars. It has a very smooth surface, much like Venus. That smooth surface is probably made of water ice. Because ice floats on water, many speculate that Europa has an ocean under its surface, and some images suggest that it may have one. However, as previously discussed, water does not mean life; it just is required for it. Remember, just because people want something to be correct, does not mean it is.

This image of Europa does make the argument for water on Europa valid (water may be deep within the cracks on the surface.) However, water alone does not mean life!

Callisto

Callisto is said to have one of the oldest surfaces of any object in the solar system, and unlike Europa, it is the most heavily bombarded object in the solar system (meaning it is not very smooth at all). Because it has so many markings from space debris slamming into it, that must mean it is very old, right? After all, the more craters something has, the longer it has been getting struck by objects. Well, if Callisto really was being struck by space debris for billions of years, it should have more small craters on its surface. This is because when an asteroid (or any type of space debris) impacts an object, it not only leaves a large marking, but it kicks up debris into the atmosphere. That debris then comes back down, and leaves behind small craters. However, with the lack of small craters, it does not seem like these were craters that came from impacts (much like Mercury's hollows). Maybe they were created, as the Bible states happened on the 4th day of Creation week. Not to mention, there is also erosion occurring on Callisto's surface, which should not be happening if it is really billions of

This image of Callisto shows that Callisto is very cratered. However, the craters that we do see do not necessarily mean age, as there should be many small craters to go along with the larger ones.

years old! This is because geological activity cannot take place on objects that are billions of years old, as all related formation activity should have ceased shortly after it formed.

Io

The last Galilean moon of Jupiter is the densest moon in the solar system. Io is known for its volcanoes, and rightfully so. The volcanoes on Io are very violent, and there are many of them. One volcano named Loki is more powerful than all of Earth's volcanoes combined! These volcanoes spew out lava frequently. Once, the size the floes from a volcanic eruption reached about as large as the state of Arizona! If Io is really billions over years old, calculations show that it would have recycled its own lava over 25 times (assuming that Io was still erupting as much as it is today.) However, if it is only several thousand years of age (as the Bible indicates), then it

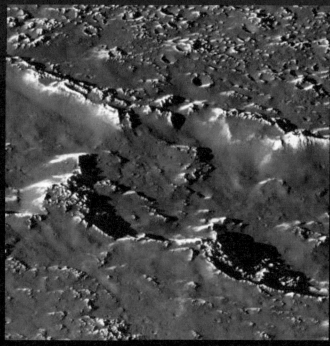

This image shows geological activity on Callisto, which should not be happening if Callisto is billions of years old!

is natural for Io to be erupting this intensely! Also, Io produces much heat within its interior, almost twice the amount of Earth's. Of course, some of the heat comes from tidal flexing, which happens when Jupiter tugs on Io from one side, and Jupiter's other large moons tug on Io from the other. However, the amount of heat produced by this moon cannot be accounted for by just tidal flexing. Again, if Io was really billions of years old, then the energy required to produce this much heat would have been lost long ago. Another problem for cosmological models is that the lava on Io is much hotter than lava on Earth. This means that, if Io was really billions of years old, these temperatures would not be possible! This is because after a few millions years or so, Io's hot lava should have formed a crust (the upper lair of a celestial body's surface) that is not very dense, as all of the high-density material would have sunk down into the interior of Io, thus leaving a low-density crust on the surface. However, there is high-density material on the surface, physical proof that Io is young.

As you can see, Jupiter and its 4 Galilean moons flaw the secular theories in such a way that not even speculation can save them! Thankfully, there is a Creator that can build such majestic planets.

This image of Io shows a violent volcanic eruption. The amount of energy required for Io to be erupting like this for billions of years cannot be comprehended by cosmological theories.

"Jupiter is the largest of all the planets, but results in *Nature* now reveal the embarrassing fact that we know next to nothing about how – or where – it formed." - Philip Ball, 'Giant Mistake' *Nature science update*, 18 November 1999

"Talk about a major embarrassment for planetary scientists. There, blazing away in the late evening sky, are Jupiter and Saturn - the gas giants that account for 93% of the solar system's planetary mass - and no one has a satisfying explanation of how they were made." - Richard A. Kerr, 'A quickie birth for Jupiters and Saturns' *Science*, Vol. 298, 29 November, 2002, 1698-9

"I don't think the existence of Jupiter would be predicted if it weren't observed." - G.W. Wetherill, *The Formation and Evolution of Planetary Systems*, 1989, p. 27, as quoted in S.R. Taylor, *Solar System Evolution: A New Perspective*, 2001. p. 205

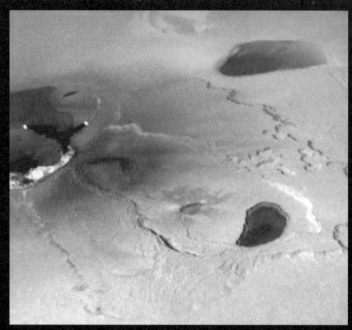

This image was caught showing us a violent volcanic eruption on Io's surface. The intensity of eruptions like this one, which occur often, shows us that Io is still very active. The only way for Io to still actively erupt like this is for it to be young, as the Bible claims.

Remember, these are the same theories that secular astronomers boast about, claiming that they are fact. As William Shakespeare said-

"A fool doth think he is wise, but the wise man knows himself to be a fool."

Saturn

When people think of Saturn, the first thing that comes to their minds is Saturn's beautiful rings. There is no doubt about it, Saturn's rings are gorgeous. These beautiful belts of icy particles make Saturn one of the most prominent signs of astronomy. So, how do people who do not believe in a Creator attempt to explain their existence? Apparently, they are not sure yet. One famous theory suggests that a moon or asteroid passed within Saturn's Roche limit. Wait, what is a "Roche limit"? Well, each planet has a Roche limit, named after the man which theorized them. Basically, if an object passes through a certain area near the planet (the Roche limit) it breaks up, due to the unstable attraction with the planet. As stated earlier, apparently a large moon or asteroid slipped passed this limit, and created the rings. Technically this is possible, but it is important to note that all 4 gas giant

While all 4 gas giant planets have rings, none of them can compare to the magnificence of Saturn's.

planets have rings (Jupiter, Uranus and Neptune along with Saturn). For Jupiter and Neptune, one of their asteroid moons could have passed within the limit, because asteroids have very uneven orbits around their planets. Maybe this happened for Uranus, as well, but surely not Saturn. Saturn's rings do not seem to consist of what asteroids typically contain, and the rings are clearly too large to come from just one asteroid.

Similarly, some people suggest that maybe a large moon of Saturn was pushed into the Roche limit, by an (you guessed it) asteroid. As was the case for the 4 terrestrial planets, the asteroid would have needed to approach Saturn at the perfect angle, and it would have needed to be a perfect size to make sure that the moon would not have been demolished, yet possess enough force to move a (presumably) gravitationally locked moon. Another theory suggests that maybe a comet came too close to Saturn, and then broke up. The chances for that, though, are extremely slim. Just like an asteroid, it would have needed to approach Saturn at such a minuscule angle, and it

Some people theorize that a large moon of Saturn, or an asteroid, was shattered by coming too close to Saturn, producing its rings. However, despite the many theories, it is still difficult to find a legitmate answer.

MΩ

If there is no dynamo for a planet's magnetic field, it should not exist! It would have surely decayed by now.

probably would have needed to make several flybys of Saturn in order to break apart. Obviously, the most rational theory would seem to be that of the Roche limit. Unfortunately for cosmologists, an asteroid was recently discovered that had rings around it, just like Saturn! What could have possibly crossed the asteroid's thin Roche limit (if it even has one) to create its rings? It makes astronomers rethink the entire Roche limit idea. Maybe these rings really did not form from asteroids, moons or comets after all.

"Yet even today, just how and when each of the rings each formed remains unknown." - http://www.space.com/7165-enduring-mystery-saturn-rings.html

Also, Saturn's magnetic field is very symmetric, meaning it is even on all sides. Therefore, Saturn's magnetic field does not seem to be coming from a dynamo. This is impossible if Saturn is even millions of years old (nevertheless billions), as Saturn's magnetic field would have decayed by then.

"Saturn dumbfounded planetary theorists who study dynamo models by having a highly symmetric internal magnetic field. A field that is symmetric about the rotation axis violates a basic theorem of magnetic dynamos." - Fran Baganel, 'A New Spin on Saturn's Rotation', *Science* 316:380, 20 April 2007

Not to mention, if the whole cloud of gas and dust theory (the one that supposedly formed **every** planet in the solar system) was true, Saturn (and Jupiter, too) should not exist! Basically, Jupiter and Saturn would have both moved into the Sun over time, and then would have slammed into it (this would have been early in the solar system's history, and thus it is debated on whether or not Newton's first law of motion would have applied here). A report on this said-

"These theories fail to describe the formation of gas giant planets in a satisfactory way. Gravitational interaction between the gaseous protoplantetary disc and the massive planetary cores causes them to move rapidly inward over about 100,000 years in what we call the 'migration' of the planet in the disc. Theories predict that the giant protoplanets will merge into the central star before planets have time to form. This makes it very difficult to understand how they can form at all." - Astronomy and Astrophysics press release *The locked migration of giant protoplanets* 21 March 2006

Moons of Saturn

Much like Jupiter, Saturn has many moons. Some, such as Tethys and Dione, are geologically active, which is practically impossible after billions of years. Others have very intriguing features as well, so let's check them out.

Titan

After Ganymede, Titan is the second largest moon in the solar system. Unlike other moons in our solar system, Titan has a thick atmosphere. This atmosphere is composed of methane and nitrogen. The methane creates a huge problem for cosmology. Because sunlight breaks methane down, Titan should not have had methane for billions of years! One study even claimed that the methane could have only lasted for about 10 million years at most. Because of this, there must be some source of methane on Titan. Though some predicted a global ocean of methane and ethane on Titan, a probe that landed on Titan's surface confirmed that there was no ocean. There are a few lakes made up of methane and ethane, but they (along with the vents on Titan) are probably not enough to account for 4.6 billion years worth of methane.

Titan is the largest moon of Saturn. This moon's atmosphere contains methane and nitrogen, both of which seemingly conflict the cosmological model. The methane cannot really be accounted for after billions of years, and the nitrogen requires speculation to fit into the secular theories.

There are some theories that state that Titan formed elsewhere in the solar system, allowing it to keep some of the elements within Titan's atmosphere. Then, as they would claim, Titan would have migrated towards Saturn, which would allow Saturn to acquire the moon as a natural satellite. Unfortunately, just like the theory that states Earth's moon could have come by such a manner, it is impossible for a celestial body to capture another like this, unless there was assistance by other astronomical bodies. And remember, they are still very speculative theories that cannot be confirmed.

Enceladus

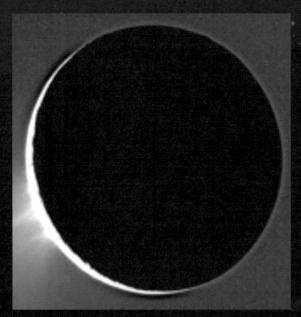

Towards the left-hand side of the moon, you can see Enceladus spraying out ice.

Enceladus is known for "spray painting" several moons of Saturn. This is because there is a large geyser that has been spewing icy material out into space, which proceeds to lands on other Saturnian moons. This should not be happening if Enceladus is billions of years old (much like Io), because an enormous amount of energy is required, which would most likely have dissipated after such a long time. Recently, though, scientists have discovered that Enceladus is tugging on the Cassini spacecraft. That may mean there is an ocean inside of Enceladus, which could account for the powerful geyser. However, just like Mars and Europa, social media claims that the ocean (which is not confirmed actually there) means life. Remember, water alone does not mean life. In fact, even if there was life on other planets or moons, it would not really help cosmologists anyways. After all, where (and how) in the universe would life be able to originate all on its own?

Janus & Epimetheus

Janus and Epimetheus are tiny asteroid moons, nicknamed the "Siamese Twins" of Saturn. That is because every 4 years, they switch places in their orbits! So if Janus is on the outside orbit, and Epimetheus is on the inside, Epimetheus will go to the outside orbit, and Janus to the inside! This has to be perfectly balanced in order for their gravitational interactions on each other to work. The odds for such a balance are extremely slim. Surely it points to intelligent design.

Truly, Saturn is absolutely marvelous. Not only are its rings and moons beautiful, but it also shows us that cosmology cannot account for such perfect balance and majesty. And sometimes, it is planets like Saturn that humble us. They show us that we humans are so little compared to the universe. And these planets make us ask ourselves one question, "What point is there to try and outsmart intelligent design?"

These two asteroid moons may not seem very interesting, but what they do is!

The bright blue dot that you see is none other than our home, Earth. Not only is our world perfectly balanced, but so are the other planets and moons in our solar system.

Uranus

Unlike the previous planets, Uranus was not considered an actual planet until the March of 1781. That is because it is barely visible with the naked eye, and also because it orbits too slowly around the Sun that astronomers of old did not know much better. Because of this, we did not know much about Uranus until Voyager 2 flew by it.

Image of the 7th planet.

When it arrived, we learned many fascinating things about what many people considered a relatively dull world. For example, Uranus' poles are tilted drastically, so while all the other planets spin like a steering wheel as they orbit the Sun, Uranus rolls like a ball. This is impossible according to secular models, because all planets are supposed to "stand" upright, and spin like a steering wheel. So how do cosmologists attempt to explain this? An asteroid collision... Is there any proof for such a collision? Not at all, in fact, there are some features of Uranus that seem unlikely if a collision occurred. First of all, Uranus is a gas giant, meaning that its surface is deep within the atmosphere, or nonexistent (excluding the core). This means that in order for an asteroid to cause so much damage, it had to be substantially large, yet of limited power (to make sure that the developing core was not destroyed). Also, Uranus orbits the Sun within the plane of the ecliptic more than any other planet besides Earth. Surely, even if a collision did occur, it would not have produced such a balance!

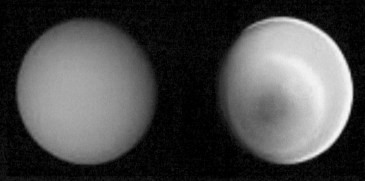

To the left is Uranus to the naked eye.

To the right is a false-color image of Uranus, showing a pole to be in an abnormal position.

That is why some people believe that maybe Uranus once was an asteroid that was the "ball in a game of ping-pong" between Jupiter and Saturn. While it was being bounced (or slung) back and forth between the planets, its poles were tilted, and after being bounced several times, it was finally shot out where it is today, when Saturn "missed the hit". If this seems absurd to you, that is because... let's just say planets do not play ping-pong. Not only does it seem very unlikely that Jupiter and Saturn could align several times to "toss" it back and forth, but it seems unlikely that Uranus could become so properly aligned in the plane of the ecliptic!

Also, Uranus' magnetic field is extremely offset from its core, and is not aligned with its spin axis. If the cloud of gas and dust model was true, and the planet's magnetic field was formed the way other planets' did, then there is no reason that such a powerful magnetic field should be so offset. Also, unlike other gas giants, Uranus does not radiate much heat into space. As you can see, just a flyby of this planet has been a huge problem for cosmological theories. But there are still more massive problems for the Enemy of the Creator's lies.

Moons of Uranus

Because Uranus' poles are positioned horizontally (where the equator should be,) the equator is positioned vertically (where the poles should be). Since the moons of Uranus orbit around the equator, they seemingly orbit up and down. That means that the moons could not have formed before a collision (a collision that is necessary in order for cosmological models to account for Uranus), or else they would not orbit where they are today. They also could not have formed after a collision, because Uranus had supposedly already formed when a collision occurred. They could not have even formed during a collision, because they have regular orbits. As one Nobel-Prize winner explains-

"To place the Uranian satellites in their present (almost coplanar circular) orbits would require all the trajectory control sophistication of modern space technology. It is unlikely that any natural phenomenon, involving bodies emitted from Uranus, could have achieved this result." - H. Alfven and G. Arrhenius, *Structure and Evolutionary History of the Solar System*, 1975, D. Reidel Publishing Co., Dordrecht, Holland and Boston, MA, p. 219

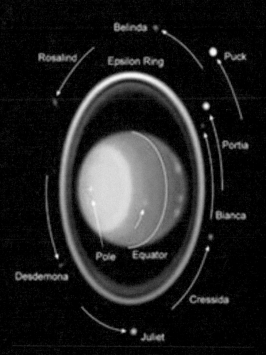

As you can see from this false-color image of Uranus, the moons of Uranus (like all other moons in the solar system) orbit around Uranus' equator (which is positioned vertically instead of horizontally).

Miranda

Miranda is the 5th largest moon of Uranus, and is very peculiar to say the least. On Miranda, the terrain drastically changes (even more so than on Ganymede.) There is smooth terrain, rugged valleys, cratered areas and hilly landscapes (keep in mind, Miranda is only about 300 miles across.) So, how do secular astronomers attempt to explain such an anomaly? Take a guess. If you guessed an asteroid, you were almost right. The correct answer is that actually, 5 asteroids slammed into Miranda! Thankfully, there is much debate about this. Most scientists agree that such a moon would not even be able to survive 1 asteroid. Therefore, it is safe to assume that Miranda could not have possibly survived 5.

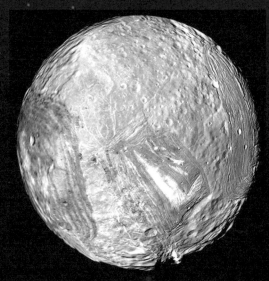

As you can see in this image, Miranda's surface features change dramatically.

"Although some sort of collisional disruption appears to be required, it is not obvious that the present terrain, with relief up to 20km, would survive catastrophic disruption and reassembly." - Taylor, *Solar System Evolution: A New Perspective*, Cambridge University Press, p. 261, 1992

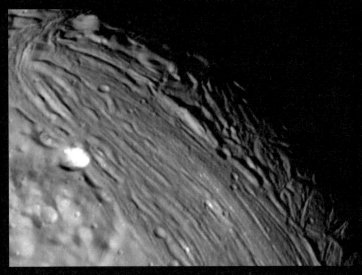

This image on Miranda shows drastic land change, from a more cratered area, to a smooth landspace, and then on to rough terrain. The best current theories cannot remotely explain how this is so on such a small celestial body.

"Miranda's appearance can be explained by theories, but the real reason is still unknown." - http://solarsystem.nasa.gov/planets/profile.cfm?Object=Miranda as of November 1 2007

"The upwilling and shattering explanations for Miranda's bizarre appearance are really just speculation. Much more evidence is needed to provide a satisfactory explanation." - http://nineplanets.org/miranda.html

Miranda is truly a mystery. It is a shame that some people are still willing to believe that 5 asteroids slammed into it, rather than give any thought to the idea of there being an Intelligent Designer. Uranus, while still being a huge mystery to us, shows us that no matter how appealing secular theories appear to be, they are little more than overrated lies.

Neptune

Neptune is the farthest known planet from the Sun. In fact, it is so far away that it was "discovered" by a mathematician who figured that due to the way Uranus orbited, there should be a planet behind it. Astronomers then pointed their most high-tech telescopes at the point where Neptune was predicted to be, and there it was. Just like Uranus, we do not know much about this gas giant. Most of the data we have is from the flyby of Voyager 2, the same probe that flew by Uranus. However, just like every other planet, what we do know does not seem to help secular astronomers fit the pieces to their puzzle at all.

First of all, just like every gas giant, Neptune has many storms. Neptune even has a storm called "Scooter," which rushes around the planet in just 16 hours! There was also a feature on Neptune called "The Great Dark Spot" which was a huge storm system that had been found in multiple places on Neptune! Neptune also has the fastest winds measured anywhere in the solar system. This means Neptune (just like all of the other gas giants) is still very active, even though it is thought to be 4.6 billion years old, which means it should be relatively quiet.

In the center of this image, you can see the famous Great Dark Spot, which has now subsided. It shows that Neptune is a dynamic, violent place that is still very active.

Also, there are magnetic field problems, again! Just like Uranus' magnetic field, Neptune's magnetic field is greatly offset from the core!

"However, as with Uranus, the Neptunian field is tilted greatly with respect to the planet's spin axis, and it is not centered on the core. The reason for this anomaly is unknown." - Mark. A. Garlick, *The illustrated Atlas of the Universe*, p.100

At the time, some astronomers believed that maybe Voyager 2 flew by Uranus just as its poles were flipping, which would explain such an offset. However, because Neptune's magnetic field was just like Uranus', it would seem unlikely that 2 planets were undergoing polarity reversals at the same time!

"It seems that the possibility of finding two planets both experiencing magnetic polarity reversals is small." - *Christiansen and Hamblin, Exploring the Planets*, p. 424

Also, it may interest you to know that Neptune (and Uranus as well) cannot exist if the whole gas and dust cloud theory is true! As Astronomy magazine explains -

"Pssst… Astronomers who model the formation of the solar system have kept a dirty little secret: Uranus and Neptune don't exist. Or at least computer simulations have never explained how planets as big as the two gas giants could form so far from the Sun. Bodies orbited so slowly in the outer parts of the Sun's protoplanetary disk that the slow process of gravitational accretion [the process in which clumps of gas and dust stick together to form the celestial bodies within our solar system] would need more time than the age of the solar system to form bodies with 14.5 and 17.1 times the mass of Earth" - R.N., Birth of Uranus and Neptune, *Astronomy*, 28(4):30, 2000

Because the planetesimals would be too far from the Sun, it seems as if it would take them way too long to produce Uranus and Neptune. As other sources state -

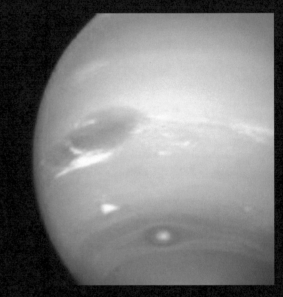

"What is clear is that simple banging together of planetesimals to construct planets takes too long in this remote outer part of the solar system. The time needed exceeds the age of the solar system. We see Uranus and Neptune, but the modest requirement that these planets exist has not been met by this model." - S.R. Taylor, *Destiny or Chance: our solar system and its place in the cosmos*, p. 73

Neptune is too far away from the Sun to have even evolved in 4.6 billion years! The time required would have been over twice the theorized age of the solar system.

"There have been many attempts to model the evolution of a swarm of colliding planetesimals… Safronov calculated the characteristic timescales for planetary growth. In the terrestrial region, he found timescales of 10^7 years, but the time estimates increased rapidly in the outer regions of the solar system, and was 10^{10} years for Neptune – which is twice the age of the solar system. It is clear that, in view of the large timescales found for the formation of the outer planets, a satisfactory theoretical model for the accretion of planets from diffuse material is not available at present." - J.R. Dormand and M.M. Woolfson, *The Origin of the Solar System: the capture theory*, p. 39

"It's clear that our level of sophistication of studying planet formation is relatively primitive... So far, it's been very difficult for anybody to come up with a scenario that actually produces Uranus and Neptune." - Martin Duncan, Queens University, quoted in *Astronomy* 28(4):30

As you can see, we know little about how these cosmological ideas are even possible! What is worse is that, these are the same theories being taught as fact to children and scientists alike! However, when they bypass what the books say, and look into the actual evidence, it is quite obvious that these secular astronomy models cannot remotely be correct. As the above quote states, "It's been very difficult for anybody to **come up with a scenario**..." This sentence shows that, not only is the cosmological theory incorrect, but it also shows us that the goal of the atheist is not just to try and model the way the universe formed. It is to try and model the way the universe formed without a Creator! It does not matter how flawed the story is. As long as it can deny the existence of their Creator, they will gladly accept it like blind sheep. However, as King Solomon said-

"Any story sounds true until someone sets the record straight." - Proverbs 18:17 NLT

Triton

Triton is the largest moon of Neptune. Once again, another moon, so astonishingly far away from us, makes the secular theory appear redundant. For starters, Triton is still very geologically active! As stated earlier, geological activity means an object must be young, and that is no different

for Triton. This activity changes the way cosmologists believe the moon became a satellite of Neptune. Some cosmologists believe that Triton is old, and the geological activity is accounted for by tidal flexing (somewhat like Io). However, it is unlikely that tidal flexing accounts for all of the geological activity on Triton, and unlike Io, it has no external source to produce heat within (besides Neptune). Also, Triton's orbit around Neptune is decaying, which means one day it will slam into Neptune. Surely Triton would have already slammed into Neptune if it was billions of years old! These reasons have led cosmologists to believe that Triton was once a TNO (Trans-Neptunian Object) that came too close to Neptune, and was caught within its gravitational pull. That would

Triton's surface is young, which it should not be if our solar system is billions of years old.

explain Triton's retrograde (backwards) orbit around Neptune, and since Triton's surface somewhat resembles that of Pluto's (another TNO), surely that is proof of how it became a moon of Neptune...

right? Unfortunately for those who believe this, studies have shown that an object cannot pull another gravitationally inwards like this, much like how Earth's moon, and Saturn's moon Titan, could not have come from such an event. Obviously, neither one of these theories work, and they never will unless secular theorists decide to accept their Creator.

This brings us to the end of Neptune, and to the end of the planets in our solar system. As you can see for yourself, this planet defies cosmological theories in the largest ways possible (it should not exist). Remember Neptune when someone tries to tell you that secular theories can explain how the universe came to be, without a Creator.

According to cosmology, all four gas giants should not exist! A crying shame that we cannot just enjoy what was put in place, rather than try and explain everything on our own.

Comets

The final "part" of the solar system we will cover is about comets. Comets used to be a symbol of fear, but nowadays we can predict when they will come, thus allowing us to enjoy their beauty, rather than ignore it. The part of the comet that is so beautiful is the "tail" that follows it. The tail forms when the comet gets too close to the Sun. That is when the materials within the comet will "melt" and be released behind it. So, where do these comets come from?

This comet (C/2001 Q4) came close to the Sun, which spawned its tail.

This is Halley's Comet, which was the first comet to have its "coming" predicted. The next time it will be visible with the naked eye is in 2061.

The answer to this question is (surprisingly) debated, but before we discuss where they come from, we should realize where comets go. First of all, comets begin their journey when they are gravitationally disturbed by the Sun or another star. Afterwards, they normally travel inward towards the Sun. They then can either crash into another object, be slung out of the solar system completely, or maybe make another lap around the solar system. However, because they do not last long (and because we still many comets today), where do the ones we see today come from? Some astronomers believe that there is a huge cloud called the Oort Cloud, about 1.5 light years away from the Sun. This is where comets come from, and it explains why we still see many today, because there are so many comet nuclei (the solid part of the comet) out there to replace the depleted comets, even after billions of years. However, there is no evidence for the existence of such a cloud. As Carl Sagan stated-

"Many scientific papers are written each year about the Oort Cloud: Its properties, its origin, its evolution. Yet there is not a shred of direct observational evidence for its existence." - Carl Sagan and Ann Druyan, Comets, 1997, p.230

Secular models cover this up by claiming that the Oort Cloud is too far away to see it. You just have to **believe it exists**... interesting. Anyways, another problem with the Oort Cloud idea is that the comet nuclei are too far away from the Sun. This is impossible, because the nuclei would have surely impacted each other while travelling that far away, thus breaking each other up, with a very few surviving the trip. Finally, some comets are known as "short-period" comets, because they orbit the Sun in 200 years or less. The maximum lifespan of a short-period comet is around 10,000 years. Obviously, the Oort Cloud (which supposedly is billions of years old) cannot seem to supply short-period comets!

Another area that is required for comets to exist after billions of years is known as the Kuiper Belt. The Kuiper Belt is an area beyond Neptune that supposedly houses many comet nuclei. Unfortunately for cosmology, the Kuiper Belt idea does not work much better than the Oort Cloud. This is because, just like the Oort Cloud, there needs to be an "endless" supply of comet nuclei to make up for the short-period comets that we see today. A few years ago, a study was done (using the Hubble Space Telescope) to confirm that these comet nuclei are really there. To the surprise of many, the astronomers who worked on this project found very few comet nuclei. Remember, the cosmological model requires millions of them to be out there.

Not only are the Oort Cloud and Kuiper Belt models rather unreliable, but we have also sent space probes to find out what these comets are actually made of. The secular theory predicts that comets can only contain certain elements (mostly volatile ones), but we have actually found many other elements that could not have been out there if the secular theory was true! The elements that were predicted to be in these comets were uncommon or not there at all. It is interesting how these "space snowballs" make such a theory appear so utterly bankrupt.

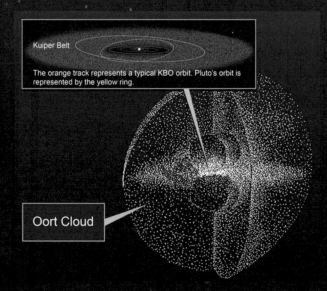

Kuiper Belt

The orange track represents a typical KBO orbit. Pluto's orbit is represented by the yellow ring.

Oort Cloud

This is an artist's conception of what the Oort Cloud (the large sphere) and Kuiper Belt (the flat oval in the top left corner) may look like. Of course, there is no physical evidence that the Oort Cloud exists, and though some objects like the dwarf-planets Pluto and Eris seem to be in the Kuiper Belt, that does not mean there are millions of comet nuclei out there with them.

When probes were sent to comets, we discovered that they do not contain the materials cosmology had predicted!

Stars

Just like the Sun and Moon, we see stars regularly, and often forget about how beautiful they are. The Bible states that stars were put in place, and in the Bible the Creator uses constellations to point things out to His followers, as He did for Job (Job 38:31).

Beyond the confines of our stellar neighborhood, we now encounter stars. Do our distant nightlights prove secular theories? Of course not! First off, stars are formed by a cloud of gas and dust swirling together and collapsing upon itself, which would then condense into the star. The problem with this idea, however, is that a cloud of gas and dust would not stay together for long at all! Gravity, of course, would hold the cloud together, but where exactly would the gravity come from? Apparently from other gas and dust particles within the cloud (before the cloud would collapse on itself to make the star). Raise your hand. Congratulations, you just broke the entire gravity of Earth tugging on your hand. How was your hand lifted? Muscles within your arms. The same is true for these clouds of gas and dust. Gravity is already a weak force, nevertheless gravity coming from minuscule dust particles. It is so weak that another "greater force" called gas pressure would cause the cloud to disperse (your muscles, respectively). As you can see, it would result in no gas cloud, which in turn would result in no star.

There is a solution to this problem, though. It is when a supernova (a dying star that releases its energy) occurs. This energy would negate the gas pressure within the area, and allow clouds of gas and dust to form during the time. However, this means that for all stars to have formed, a supernova must have needed to occur. Since a supernova is a dying star, it does not explain how the first stars could have formed on their own, because unless there was an infinite loop of stars (which even secular scientists

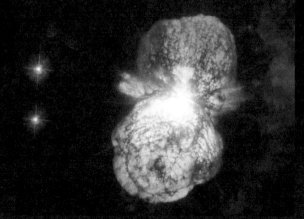

This is an image of an upcoming supernova, which would then enable star formation in the surrounding region. However, supernovae do not explain where the first stars came from, because a supernova (which requires a star) must occur to form stars!

agree there could not have been, due to an expanding universe) there would have been no supernova for the first stars to have used!

Also, we have previously discussed on how a fair number of stars are binary (when 2 or more stars orbit each other). This poses the question - if somehow clouds of gas and dust could actually turn into stars and planets, how would 2 stars be able to orbit each other to form? This has been a huge puzzle for secular astronomers, because all possibilities that such an anomaly could occur are slim. First, a supernova must occur. Then, 2 clouds of gas and dust need to condense into a star at the exact same time, because if one star formed before the other, it would tear the other cloud of gas and dust apart! And because of the presence of the other star, even if it was somehow possible, the planetesimals could not condense into planets or moons, lest they be torn apart by the opposing star's gravity.

This is an artist's conception of a binary star system.

"Literally hundreds of ideas on how stars are formed have been advanced in past decades. However, we are still far from any real solution." - Martin Harwit, *Astrophysical Concepts*, Second Edition, p. 405

"Nobody really understands how star formation proceeds. It's really remarkable." - Roger A. Windhorst, as quoted by Corey S. Powell, "A Matter of Timing" *Scientific American*, Vol. 267, October 1992, p.30

"The universe we see when we look out to its furthest horizons contains a hundred billion galaxies. Each of these galaxies contains another hundred billion stars. That's 10^{22} stars all told. The silent embarrassment of modern astrophysics is that we do not know how even a single one of these stars managed to form. There's no lack of ideas, of course. We just can't substantiate them." - Martin Harwit, Book Review in *Science*, 7 March 1996, p. 1201-2

Interestingly enough, some cosmologists will tell you that we actually see stars being formed within clouds of gas and dust. However, even if stars could form by a cloud of gas and dust, it is

This is suppoesdly an area where stars are forming. Whether stars form here or not, we cannot be sure, because we cannot actually see them!

impossible to know that new stars are actually what we see within those clouds. This is because stars would be forming in the center of a cloud of gas and dust, thus it should be obscured by the light of the surrounding debris! And while some may claim that we can see stars being born by using different light frequencies, the light given off by a newly-formed star should also be the same as the gases that make up the star (once more, making assumptions less credible.) Not to mention, us humans have been observing stars with such detail for only a few decades, so it is very difficult to clarify anything we see.

Also, stars are located within galaxies. Galaxies are very strange, and difficult to manage for the secular model. First of all, if you get a cup of cold water, and leave it outside on a hot, sunny day, after a while the cold water will become warm. The air would also get cooler, but by a very, very small amount. This is also true for the universe. Space itself is thought to be cool; however, galaxies seem to be very hot. The Big Bang Theory (the theory that states that a huge "explosion" out of nowhere formed the universe) states that the "explosion" would have been very smooth. Instead, we see such drastic temperature changes within the universe. After billions of years (over 13 billion for the entire universe) the universe should have cooled/warmed to the same general temperature and not experience such differences. In speaking of which, in some areas of space we see many

In this image, you can see an entire cluster of galaxies in one area. However, in many areas across the visible universe, there are few galaxies, or none at all. Like any other explosion, debris should more or less be evenly spread, but that does not seem to be the case if a Big Bang occurred.

galaxies, while in others, we see nothing but empty space. Once again, the Big Bang Theory states that space should be very smooth and almost even, but apparently that is not what we see!

"...we would not give very high odds that any of these theories is a useful approximation of how galaxies actually formed." - Joseph Silk, (Head of Astrophysics at Oxford) and P.J.E. Peebles (Albert Einstein Professor of Science Emeritus at Princeton), "A Cosmic Book of Phenomena", *Nature* 346 233-239

As you can see, no star or galaxy should even exist. Thankfully though, there is a Creator who could make such glorious objects.

"When I look at the night sky and see the work of Your fingers-the moon and the stars you have set in place- what are mortals that you should think of us, mere humans that you should care for us?" - Psalms 8:3-4 NLT

Epilogue

In the second paragraph of this book, I stated that I would tell you which theory was correct and why. However, I told you which theory was incorrect instead. Indeed, it is difficult to find proof for a Creator in astronomy, but actually, I think you have seen for yourself the perfect balance of the entire universe. Of course, though, there are many artifacts mentioned in the Bible that have been discovered, history matches what the Bible claims perfectly, and what the Bible stated would come true has, or is right now. That is the reason for this novel. You have been told by others that the cosmological model is true, just as I have. Little do they know, however, that they are playing right into the foretelling of the Holy Book.

"For a time is coming when people will no longer listen to right teaching. They will follow their own desires and will look for teachers who will tell them whatever they want to hear. They will reject the truth and follow strange myths." -2 Timothy 4:3-4 NLT

The most unfortunate thing of all is that many atheists will reject all of this. They will act as if the Bible is an outdated lie, yet some cannot even defend their arguments. Just because their parents and schoolbooks told them that a "Big Bang" occurred, they will blindly believe it. This is because they do not want to believe there is a Creator, and that there will be a Day of Judgment. Every person will be judged by what they have done, and because many of those who reject the truth love physical pleasure (which they know they would be punished for), they just simply say "It will not happen." In fact, some even scoff at those who believe the Bible, just because they are that insecure. This is a tactic that dates back to the ancient Greeks, though, when they knew that they had lost an argument, and would then proceed to belittle their opponent for self-pride.

This is especially true, both in the past, and today. In the past, there were theories similar to the secular models that people took very seriously, just because they wanted to be right. However, when other people could show them their flaws, some would actually get violent! Of course, they were wrong, but their pride blinded them to the truth. Today, there are thousands of degreed scientists who believe in a Creator, yet they are mocked by others constantly. It would seem as though their fellow men and women would have something better to do, but instead they constantly slander them.

Of course, not all atheists are like this. And if you are an atheist who is reading this, do not fall for the lies of this world. Remember -

"What is popular is not always right, but what is right is not always popular." - Albert Einstein

Yes, science is always changing, and it may have even changed by the time you are reading this! However, do not let anyone make up your mind. Just realize that the current secular models of astronomy are flawed, and right now there are more questions than solutions. In fact, we humans know so comparatively little about space that, thanks to recent discoveries, we are not even certain about how many planets are in our solar system. Remember, too, that there will be a Day of Judgment. I am certain you do not want to spend an eternity in flames, though there is hope. In fact, the Bible is a message of liberation, not condemnation. The Creator of the Universe (God) wants you to be His very own child, so He can show you marvelous things. He wants you to be in His Kingdom, so that you may live forever.

By the way, not only do thousands of scientists accept the idea of a Creator, but some of the greatest scientific minds in history gave light to Creationism as well.

"This most beautiful system of the sun, planets and comets, could only proceed from the counsel and dominion of an intelligent and powerful Being." - Isaac Newton, *The Principia: Mathematical Principles of Natural Philosophy*

"The more I study nature, the more I stand amazed at the work of the Creator: Science brings men closer to God." - Louis Pasteur

"The more I study science, the more I believe in God." - Albert Einstein

"God [is] the author of the universe, and the free establisher of the laws of motion." - Robert Boyle

"I do not feel obliged to believe that the same God who has endowed us with sense, reason, and intellect has intended us to forgo their use." - Galileo Galilei

"I believe that the more thoroughly science is studied, the further does it take us from anything comparable to atheism. If you study science deep enough and long enough, it will force you to believe in God. - Lord William Kelvin

"It was not by accident that the greatest thinkers of all ages were deeply religious souls." - Max Planck

And for those of you who are reading that do know the Creator as your personal Savior, now you know that truly, the skies are His. Every time you gaze up into the Heavens, you can see His majesty. The Enemy of the Creator wants you to think that there is no God, so that you cannot marvel at His glory. But do not fall for his lies! If he can get you to deny the first chapter of the

Bible, he can get you to deny the plan of salvation that is discussed thereafter. Remember that the Creator of the Universe does not debate His existence, but He declares it!

"The heavens tell the glory of God. The skies display His marvelous craftsmanship. Day after day they continue to speak; night after night they make Him known. They speak without a sound or a word; their voice is silent in the skies; yet their message has gone out to all the earth, and their words to all the world." - Psalms 19:1-4a NLT

Acknowledgements

A huge thanks to Spike Psarris. His clarity on the topic of Evolution vs. Creation has helped many people, including myself. He also made me love astronomy even more than I did before I saw his work!

Special thanks to **all** of those who supported me during the making of this novel. Your kind words and time have not gone unnoticed. Without your curiosity, this novel would serve no purpose.

Special thanks to Maurice Broaddus, Jennifer Morris, and all other members who worked with me from Westbow Press. Without their patience to work with a minor, this novel would not exist.

One of the final sentences of the novel, "He does not debate His existence, but He declares it!" was quoted by Pastor John Hagee during one of his sermons. This particular phrase has been a cornerstone for me during the research and creation of this novel.

Most information comes from Spike Psarris' "What you aren't being told about Astronomy Volume 1 Our Created Solar System" and "What you aren't being told about Astronomy Volume 2 Our Created Stars and Galaxies", along with select word phrases/structures from these projects. All points (aside from the ones mentioned below) and many quotes mentioned within the novel were found by him. Other information comes from: Don DeYoung regarding information on the moon, including its eclipses, origin and implications on life. Kent Hovind regarding how the ocean was required for life to evolve. Pastor John Hagee's "Four Blood Moons" sermon, discussing how lunar/solar eclipses both are symbols for human beings. Common apologetic knowledge from www.creationastronomy.com regarding the Uranus "pinball" theory, and how atheists take their arguments seriously because they refuse to be wrong in the Epilogue. The definition of "magnetic field" was paraphrased from http://dictionary.reference.com/browse/magnetic+field Information regarding the "jelly donut" rock/beam of light found on Mars, most facts about mentioned Saturnian moons, and Venus' retrograde rotation is from NASA©, along with information explaining speculative catastrophes that are the only way to explain our moon, the lack of water on Mars and Saturn's rings. Information regarding Mercury's composition is from http://www.space.com/18643-mercury-compositon/html

Information regarding recently discovered volatile formations on Mercury is from www.creationrevolution.com. Information regarding what Mercury's name implies in Greek Mythology is from http://ancienthistory.about.com/cs/grecoromanmyth1/a/hermes/.htm Information regarding solar flare energy levels is from http://www.astronomy.com/news-observing/news/2006/11/brilliant%20flare%20seen%20%in%20ii%20pegasi Information regarding Ganymede is from www.creation.com Information regarding Io's density is from http://www.guide-to-the-universe.com/the-moon-io.html Information regarding Venus' retrograde rotation is from http://www.universetoday.com/14299/retrograde-rotation-of-venus/ Information regarding carbon dating and fossils is from Lee Strobel's *The Case for a Creator*. Information regarding Titan's methane lasting for millions of years is from Lorenz et. al, 'Titan's inventory of organic surface materials,' *Geographic Research Letters,* Vol. 35, page LO2206, January 29, 2008. Information regarding thousands of scientists accepting Creation can be found on www.rae.org/darwinskeptics.html (as quoted during Spike Psarris' What you aren't being told about Astronomy: Volume 1 Our Created Solar System). Information regarding the moon's albedo is from http://www.universetoday.com/75891/why-does-the-moon-shine/ Information regarding Tethys' and Dione's geological activity is from Moons of Saturn, Uranus and Neptune http://lasp.colorado.edu/education/outerplanets/moons op.php Information regarding on how Saturn's rings seem too spectacular compared to other gas giants that they probably did not form the same way as theirs if found from http://www.space.com/7165-enduring-mystery-saturn-rings.html Information regarding Roche limits is from http://www.teachastronomy.com/astropedia/article/The-Roche-Limit Information regarding comets being a symbol of fear in ancient times comes from http://deepimpact.umd.edu/science/comets-cultures.html Information regarding when Halley's Comet will be visible from Earth again is from http://.curious.astro.cornell.edu/question.php?number=605 Information regarding how Halley's Comet was the first comet to have its coming predicted is from http://www.britannica.com/EBchecked/topic/252831/Halleys-Comet Information regarding how not all terrain that seems to have come from water on Mars could have come volcano eruptions and dust storms come from Mark A. Garlick's *The illustrated Atlas of the Universe* p. 63. Mentioned basic science facts (such as all gas giants have rings) can all be found on www.nasa.gov or www.jpl.nasa.gov. The fact on how for hundreds of years, people have suspected life on Mars can be found on http://spaceplace.nasa.gov/review/dr-marc-solar-system/life-on-mars.html Information regarding what seems to be a discovered Biblical artifcact is from http:///www.usatoday.com/story/news/world/2013/03/30/shroud-turin-display/2038295/ Information regarding the possibility of other planets in our solar system is from http://arstechnica.com/science/2015/01/the-solar-system-may-have-two-undiscovered-planets/ Information regarding the types of stars can be seen on an H-R (Hertzsprung-Russell) diagram. Information regarding the name of C/2001 Q4 can be found on http://solarsystem.nasa.gov/planets/profile.cfm?Object=Comets Information regarding how Neptune was discovered is from http://wanttoknowit.com/who-discovered-neptune/ Information regarding an old theory many took seriously can be found on http://www.universetoday.com/32607/geocentric-model Information regarding muscles within your arm allowing you to raise your hand is from http://www.answers.com/Q/How do muscles in your hand work Not all phrases used within the novel were thought of by me.

Prologue images: NASA – Birth of an Earth-like Planet http://www.nasa.gov/mission pages/ spitzer/multimedia/pia09931 prt.htm Solar System: Exploration: : Planets http://solarsystem. nasa.gov/planets/

Sun images: NASA – Hot Stuff http://www.nasa.gov/multimedia/imagegallery/ image feature 221.html NASA-Sunrise | NASA http://www.nasa.gov/multimedia/imagegallery/ image feature 2047.html

Mercury images: Mercury As Never Seen Before – NASA Science http://science.nasa.gov/ science-news/science-at-nasa/2008/07oct firstresults/ solarsystem.nasa.gov http://solarsystem. nasa.gov/multimedia/gallery/messenger 20081009 smooth.jpg NASA- Locations of energetic electron events relative to Mercury's magnetic field http://www.nasa.gov/missions pages/messenger/ multimedia/messenger news20110616 images4.html

Venus images: NASA – Venus Weather Not Boring After All, NASA/International Study Shows http://www.nasa.gov/topics/solarsystem/features/venus-temp20110926.html Catalog Page for PIA00104 http://photojournal.jpl.nasa.gov/catalog/PIA00104

Earth images: NASA- Public Invited to Two Free Earth Day 2012 Events at NASA Goddard http://www.nasa.gov/centers/goddard/news/features/2012/earth-day-events.html NASA- Electric Moon Jolts the Solar Wind http://www.nasa.gov/topics/solarsystem/features/electric-moon.html Photo-iss013e78295 http://spaceflight.nasa.gov/gallery/images/station/crew-13/html/iss013e78295. html Auroras | NASA http://www.nasa.gov/mission pages/sunearth/news/gallery/aurora-index. html NASA- Powerful Pixels: Mapping the "Apollo Zone" http://www.nasa.gov/topics/solarsystem/ features/apollo-zone-map prt.htm NASA GISS: Research Features: Earth's Temperature Tracker http://www.giss.nasa.gov/research/features/200711 temptracker/ NASA- Composite Image of Solar Eclipse | NASA http://www.nasa.gov/topics/solarsystem/sunearthsystem/main/News071510- Eclipse-composite.html NASA- NASA Scientists Theorize Final Growth Spurt for Planets http:// www.nasa.gov/topics/solarsystem/features/planet growth spurt.html

Mars images: Solar System Exploration: Multimedia: Gallery: Planetary Images: Mars Atmosphere http://solarsystem.nasa.gov/multimedia/display.cfm?Category=Planets&IM ID18307 NASA – Northern Ice Cap of Mars http://www.nasa.gov/missions pages/MRO/multimedia/ pia13163.html JPL | News |Years of Observing Combined Into Best-Yet Look at Mars Canyon –http:// www.jpl.nasa.gov/news/news.php?release=2006-035 APOD:2014 January 2014 – Jelly Donut Shaped Rock Appears on Mars http://apod.nasa.gov/apod/ap140129.html

Jupiter images: NASA - Covering Jupiter from Earth and Space http://www.nasa.gov/ missions pages/juno/multimedia/pia14411.html JPL | Space Images | Jupiter http://www.jpl. nasa.gov/spaceimages/details.php?id=PIA01324 Solar System Exploration: Planets: Jupiter: Moons: Ganymede: Overview http://solarsystem.nasa.gov/planets/profile.cfm?Object=Jup Ganymede Images of Ganymede http://photojournal.jpl.nasa.gov/target/Ganymede APOD: 2009 March 8 – Gibbous Europa http://apod.nasa.gov/apod/ap090308.html Images of Callisto http://photojournal. jpl.nasa.gov/target/Callisto quest.nasa.gov http://quest.nasa.gov/galileo/images/48124.jpg Images of Io http://photojournal.jpl.nasa.gov/target/Io?subselect=Mission:Voyager: APOD:2000 June 6 – A Continuous Eruption on Jupiter's Moon Io http://apod.nasa.gov/apod/ap000606.html

Saturn images: Solar System Exploration: Planets: Saturn: Gallery http://solarsystem.nasa. gov/planets/profile.cfm?Object=Saturn&Display=Gallery&Page=8 JPL | Space Images| Colorful Colossuses and Changing Hues http://www.jpl.nasa.gov/spaceimages/details.php?id=PIA14922 Cassini Solstice Mission: Saturn's magnetic field lines http://saturn.jpl.nasa.gov/photos/imagedetails/ index.cfm?imageId=1861 Solar System Exploration: News & Events: News Archive: Titan's Building Blocks Might Pre-date Saturn http://solarsystem.nasa.gov/news/display.cfm?News ID=47675 NASA – Saturn's Moon Enceladus Spreads its Influence http://www.nasa.gov/mission pages/ cassini/whycassini/cassini20110921.html Cassini Solstice Mission: The Dancing Moons http:// saturn.jpl.nasa.gov/photos/imagedetails/index.cfm?imageId=2103 Solar System Exploration: Planets: Saturn: Overview http://solarsystem.nasa.gov/planets/profile.cfm?Object=Saturn

Uranus images: Images of Uranus http://photojournal.jpl.nasa.gov/target/Uranus Space Images Wallpaper Search – NASA Jet Propulsion Laboratory http://www.jpl.nasa.gov/spaceimages/searchwp. php?category=uranus Solar System Exploration: Planets: Uranus: Moons http://solarsystem.nasa. gov/planets/profile.cfm?Display=Moons&Object=Uranus Space Images: Wallpaper – NASA Jet Propulsion Laboratory http://www.jpl.nasa.gov/spaceimages/wallpaper.php?id=PIA18185 Solar System Exploration: Planets: Uranus: Moons: Miranda: Overview http://solarsystem.nasa.gov/ planets/profile.cfm?Object=Ura Miranda

Neptune images: Solar System Exploration: Planets: Neptune: Overview http://solarsystem. nasa.gov/planets/profile.cfm?Object=Neptune Voyager – Images – Neptune Images http:// voyager.jpl.nasa.gov/image/neptune.html Solar System Exploration: Planets: Neptune: Moons http://solarsystem.nasa.gov/planets/profile.cfm?Object=Neptune&Display=Moons Solar System Exploration: Planets: Neptune: Overview http://solarsystem.nasa.gov/planets/profile. cfm?Object=Neptune

Comets images: Solar System Exploration: Planets: Comets: Overview http://solarsystem.nasa. gov/planets/profile.cfm?Object=Comets NASA – Eta Aquarid Meteor Shower: 'Up All Night' With NASA! http://www.nasa.gov/connect/chat/aquarids2011.html Solar System Exploration: Multimedia: Gallery: Planetary Images: Oort Cloud http://solarsystem.nasa.gov/multimedia/

display.cfm?Category=Planets&IM ID10195 Solar System Exploration: Multimedia: Gallery: Planetary Images: Approaching Wild 2 http://solarsystem.nasa.gov/multimedia/display. cfm?Category=Planets&IM ID=509

Stars images: APOD: 2008 March 12 – Star Forming Region LH 95 http://apod.nasa.gov/apod/ ap080312.html NASA – Preview of a Forthcoming Supernova http://www.nasa.gov/multimedia/ imagegallery/image feature 2183.html NASA – X-ray Satellites Monitor the Clashing Winds of a Colossal Binary http://www.nasa.gov/missions pages/swift/bursts/binary-clash.html The James Webb Space Telescope http://www.jwst.nasa.gov/birth.html APOD: 2009 November 21 – NGC 253: Dusty Island Universe http://apod.nasa.gov/apod/ap091121.html Solar System Exploration: Multimedia: Gallery: Planetary Images: Hubble Ultra Deep Field 2014 http://solarsystem.nasa. gov/multimedia/display.cfm?Category=Planets&IM ID=19408

Epilogue image: APOD: 2007 March 30 – Three Galaxies and a Comet http://apod.nasa.gov/ apod/ap070330.html

Printed in the United States
By Bookmasters